# IMPLEMENTING
# FREE WILL

## ADAM ALLARD

# Implementing Free Will

Copyright © 2013 by Adam Allard

Cover photo by Texelart

ISBN-13: 978-0-9896663-1-2
ISBN-10: 098966631X

Book Website
www.ImplementingFreeWill.com

Give feedback on the book at:
feedback@ImplementingFreeWill.com

Printed in U.S.A

Written on an Apple MacBook Pro using Microsoft Word and Adobe InDesign

For Sharon


and in memory of
**Erik Bartleson**
(1963-2012)

# Contents

# Preface

The topic of free will has portions of the scientific community buzzing with debates. New data suggests that we have little or no actual control over our thoughts and decisions. This is predicated on key observations about the human mind, and specifically how our thoughts seem to appear out of nowhere. This observation poses a unique problem in that it suggests on the surface that humans are not consciously directing their thoughts, and are therefore unable to actually have direct control over their free will.

While I agree with the underpinnings of these observations, as it does in fact appear that ideas, thoughts, and decisions come into our minds from nowhere, there is more to the story... much more.

I appreciate these types of discussions as they open the door of scientific inquiry, and in this case may ultimately lead to a better and deeper understanding of how the human brain operates, and how decisions are actually made. Being aware of these basic principles will then shed some insight into how we can leverage that knowledge to better direct and enhance our intellectual growth. Starting with the foundational concept that our conscious mind is not making our decisions, this book takes steps to outline where those decisions *are* being made, and why the decision "engine" in your mind works as it does. By understanding this process, we will see how we do, in fact, steer, hone, adjust, and alter future decisions, values, beliefs, and more.

This book is not meant to promote a final, end-all answer as to how the brain processes data and makes decisions, nor should it be. Rather, the intent is to provide a new starting point for discussion and analysis, opening yet another door for further inquiry and understanding.

It is my hope that having a better grasp of how and why people think the way that they do, along with outlining a roadmap that can be used for personal growth, will enhance the human experience, bring down barriers between people and cultures, and allow us all to move forward with a unified goal of making the world a better place for everyone.

If nothing else, my hope is that readers will embrace one of the book's core concepts being that open-mindedness and education is the cornerstone for solving many of humanity's problems.

I hope you enjoy reading this book as much

as I have writing it.    Throughout the process I have resisted the temptation to add superfluous information, or "fluff," for the sake of increasing page count.    In fact, I have gone back and removed a significant portion of the content, and re-written many areas in order to condense it into, hopefully, more salient and concise ideas and terminology. And In keeping with today's busy schedules, being able to deliver an idea quickly and succinctly is both important and necessary.

Enjoy your journey...

Adam Allard
June, 2013

# Implementing Free Will

Have you ever wanted to know exactly how much control you have over your thoughts and decisions? Initially you may be tempted to say, "Well of course I'm in control of my thoughts." We all seem to start with that assumption.

But if you sit down, in a quiet place with few distractions, and pay close attention to what you are actually thinking, moment-to-moment, you may be surprised at what you discover. Thoughts simply appear in your mind. Decisions seem to be made out of thin air. Why did you think what you did a few moments ago? What made that particular song pop into your head? Why did you decide to wear that type of shirt today?

And upon closer examination, you discover that, stranger yet; your conscious mind appears to be the observer. Consciousness is what acknowledges and thinks about the certain thought that just appeared, or pays attention to the song running in the background of your mind. And if your conscious mind is the observer, who, or what, is pulling the strings behind the scene?

After further thinking about a simple experiment like this, you may be led to believe that you have no control over what pops into your head. Taken further, you may begin to believe that the concept of free will, or your ability to consciously make the choices and decisions in your life is really an illusion, and that you actually have no free will.

Are we simply robots controlled by some unknown master? Depending on your background and beliefs, you may be tempted to say that your soul is guiding you, or that perhaps this phenomenon is some sort of divine intervention. Is voodoo involved in some way?

Taken on the surface, the mystery of where your thoughts emanate could be confused with a variety of possible sources. Happily there is an easy answer once you understand the process and mechanisms that are in place.

Now you may be tempted to say, "Who cares? This is all neat and semi-interesting, but obviously I've gotten along this far in my life without contemplating why green is my favorite color, or what made me decide to wear boots today, versus loafers."

Good point. But as you will see, the same mechanism that controls benign, inconsequential decisions like these also controls your fears, insecurities, values and beliefs. If growth beyond where you are today is important to you, having an understanding of the mechanisms that are governing your thoughts becomes increasingly important. Have you ever wondered why you started something and didn't finish it? Or why some people seem to be living an active exciting life while you seem to be stuck in a rut? The answer is more than just time and money. In fact those things are rarely the actual root cause of a lack of inertia. Rather they end up being yet another symptom or consequence of a basic lack of control of your thoughts.

Actually having the conscious ability to analyze, choose, direct, or change your direction is vital to personal growth, accomplishing your goals, and maximizing whatever you choose to do in the future.

Removing the mystery from the barriers in your life is the cornerstone of change. Implementing free will is not just the idea of making simple choices, but in a larger context dictates the roadmap of your life. Understanding and controlling your thoughts, reevaluating your insecurities, confronting and overcoming your fears, is both liberating and exciting. Removing these and other constraints can relegate them to a distant memory, causing you to wonder how and why you ever allowed those things limit you in the past.

This book will outline the biological reasons why your thoughts are indeed your own, even though your rational conscious mind seems to be an observer to the process. The truth is, your conscious mind is both an observer, and a director.

The theories presented in this book are designed to give you a glimpse into the vastly complex workings of the human mind, but in a simplified and easy-to-understand way. Through the various analogies and examples of how the brain works, you will be able to see that your thoughts and decisions, the essence of free will, are in fact under your control.

This book will also show that you can direct choice and free will, but only in a future tense, after you have consciously reviewed and contemplated your historical data... essentially a lagged temporal adjustment.

# Basics

To understand free will and the ability to make choices and decisions, we must first understand the basics of how the human brain operates. And while there are many different functional areas of the brain, in the context of this book we will only concern ourselves with basic data flow, storage, data manipulation, and the results that are retrieved from memory.

To better understand the mechanics of our brain, it is easier to use an analogy, such as a computer. While not a perfect analogy, computers and their systems and software can at least be a bridge of rudimentary understanding between a commonly known object and the complex workings of the human brain.

Computers are comprised of both hardware and software. The hardware is the physical components such as the central processing unit (CPU), random access memory (RAM), hard drives, and various input and output devices.

Software and data is what transforms an otherwise worthless pile of metal and plastic into a machine that has tremendous value. Both are vital to the equation, but arguably software and data, essentially the content of what you can interact with, is really where most of a computer's value comes from. The physical human brain is analogous to the computer hardware, where your conscious and subconscious minds are like the software. Data is then the knowledge and memories acquired through living.

To further narrow the scope of the computer-brain analogy, we will use a computer database as the baseline of our reference. In simplified terms, a database is a software program that stores and retrieves data. In practical terms, they are much more, in that data is usually organized and stored in a structured way. Data within the database can also be contextually connected together using "keys" or common links.

Once data has been stored into a database, it is often retrieved by way of a structured query language (SQL). This computer language directs the database to retrieve data from certain tables and can also be used to narrow the scope of the results. In other words, a database table may contain all of the names and phone numbers of everyone in your city, but instead of retrieving the entire list, the query can be constructed to only retrieve people who's last name begins with "S" or could be narrowed further to just retrieve information about a one person, or single record in the database.

Before computers, the card catalog in the library played the part of a database, although in a very simplistic way. A manual filing system with file cabinets and file folders is also analogous to a database in that information is organized and stored for later retrieval and use.

The main point to remember is that any of those databases are only as valuable as the data they contain. A library without books, or a file cabinet with no files, or a computer database without records is essentially worthless. It is important to understand the data must also be organized as well as stored, otherwise it cannot be located and utilized.

And to extend that idea, especially in the case of a computer database, they are only as valuable as the efficiency at which that data can be queried, linked, and retrieved in a meaningful format. This linking of data or connecting one point of data to another is an important, yet simple concept to understand. As an example, if you are driving across town and notice that the fuel gauge in your vehicle is indicating near empty, that is meaningful data. But simultaneously also noticing on the GPS that a gas station is only two blocks to the east demonstrates that linking two related data points can be much more valuable than one.

This may seem intuitively obvious in a simple example like this, but it is surprising the number of people who appear to fail in applying the same type of data connections in other aspects of their lives.

*What this suggests is that the value of data, however simple or complex, is generally increased when it is linked with other contextually relevant supporting data.*

Another important fact to understand is that all computer databases are only able to operate on historical data. No matter how fast a computer, it can only import data, process and store it at a given finite rate. And only after that data is stored appropriately, can it be queried and used in a meaningful way.

As a point of illustration, take Google as an example and examine how the Google database system works. At a high level, Google has written a software system that "crawls" the web, retrieving and "reading" most of the web pages on the Internet. Once each page is retrieved and parsed, the content data from that page is stored in a database. This is an ongoing, continuous process, that runs in the background, and is not something a user of Google is aware of... it is simply an automated data gathering system.

Later, you, as a Google user, visit their web page and type in a query in the form of one or more keywords. At this point, when you click the "Google Search" button or hit the enter key, Google's database engine goes to work matching the most relevant results with your query. Finally, you are presented with a page of possible answers related to your search.

Thankfully, as a user, you are not required to understand the query statements that were automatically generated in the background using your keywords, nor were you required to understand the table structure of their databases, the data keys that link one table to another, or their PageRank algorithms that order the result set into the highest probability of what you were seeking. The complexity of parsing and analyzing the data was done for you, on-the-fly, at Google's data center, in a clean, fast and efficient manner... hence the secret to their success.

Another important aspect in how well Google performs is the speed and efficiency at which your search results are returned. Some of this can be attributed to "cache" (pronounced 'cash') or holding frequently used data in computer memory, rather than retrieving it from the hard disks residing on the many servers in their facilities.

Caching is faster because data is held electrically in random access memory (RAM), compared with longer-term permanent storage on a hard drive. To retrieve data from a hard drive requires the drive to physically move a tiny arm over a sector of the disk, then reading the data from the disk, governed by the speed at which the disk is spinning, and then processing that data in some meaningful way. The time that it takes to access random data from a hard drive is rated in milliseconds, where RAM memory access is rated in nanoseconds. This represents a speed increase of up to 100,000 times faster. We will see later why your subconscious brain also needs rapid access and speed to ensure your survival.

Finally, it should be noted that computer architects and engineers did not intentionally mimic brain function in their original designs, but rather their designs, driven by the need for speed and efficiency, have, accidentally perhaps, evolved into systems that are becoming closer to analogous with brain function. And while this book does not care about computer system architecture, a simplified version of that topic, as outlined above, will serve as a good reference analogy for brain function in the coming pages, especially as it relates to gathering, processing, storage and retrieval of data, as well as the higher order functions of thoughts, decisions, values, morals and emotion.

# Information Gathering and Storage

It has been suggested that the brain stores information holographically [1]. A true holograph is created from a coherent light source, such as a laser, and is formed by splitting the beam into two paths. One beam is then directed onto photographic film, and the other illuminates the object. The resulting "image" captured on the film is actually just an interference pattern between the direct light source and the light reflected from the object. The interesting part of all of this is that unlike a regular photograph, if you cut a holographic film in half, both pieces still contain the entire picture, and only the resolution is reduced. This holds true no matter how many pieces are cut.

In essence, every part contains the whole, just individually at a lower resolution. And while the human brain obviously does not store information with a laser, the holographic analogy is a good one in terms of data being replicated, spread and stored across regions rather than in one particular spot. This holographic analogy also explains why people who have suffered brain damage can sometimes regain their memory because even though one area of their brain was affected, the other unaffected areas still contain "copies" of the data [2].

During the course of living your life, your brain is bombarded with data. Every waking moment is full of sights, sounds, smells, and tactical feedback from nerve endings all over your body. All of that incoming data into the brain is processed and temporarily stored.

The trigger for data processing, however, is an interesting area; in that there are filters our brain applies to limit an "overload" of data. Think about how much data your brain is taking in at this very moment... touch, smell, sight, sound. If unfiltered, you would be paralyzed in trying to process what just happened in the last thirty seconds. This filtering concept or conversely the "analysis paralysis" arising from a lack of it, is an important concept in understanding how and why the brain functions as it does.

These data filters can be thought of as an exception-based system. Many animals, humans included, appear to operate on an exception basis when it comes to information processing. Domestic house cats, for example, are keen on noticing differences in their environment. For those of you who have cats, you are probably aware that if you change the arrangement of furniture, or place a box on the floor, your cat will, upon entering later, immediately notice and "cat–alog", as I like to call it, that change. In other words, an exception to what was normal before has occurred.

Once the exception or change has been explored and noted, the cat continues on its way and incorporates those changes as the new "normal." This biologic programming is also applied to dynamic environmental changes in scenery, and is the mechanism by which they notice prey through movement.

The human brain, like the domestic cat brain, is tasked with many functions, some automatic, others responsive. Noticing change is an important aspect of how the human mind works, however we take the implications to completely new levels, compared to the domestic cat, by using it as a filter to determine what is important for the conscious mind to contemplate and reflect upon.

Interesting side note: want to know why you can take an entire class on something and years later barely remember anything about it? More than likely your retention ability is directly related to the importance you placed on the topic being covered (i.e. you are more likely to remember details about things in which you are interested, rather than not)... more about this later.

All of these data gathering, filtering, and storage tasks take place in our brains continually, with most data being relegated to inconsequential, and therefore a low-degree of resolution in terms of conscious thought or storage retention.

*Data is the inventory of tools your conscious mind uses to construct all that you know, feel, value, and believe.*

# The Conscious Data Manager

Rather than being involved with making decisions in real time, the purpose of the conscious mind is to reason and rationalize using historical data. Essentially, consciousness is really about parsing, cleaning, linking, and storing updated (rationalized) historical data that our subconscious (cache and operating system) can use for rapid decision-making and system operations.

Consciousness is really a background processing system. Although it plays a crucial part of formulating the basis for decisions, its role is primarily for contemplating historical data. This is the primary reason why thoughts that appear in your head seem to come from somewhere else, because it is only after they appear that the conscious mind can contemplate them. Consciousness is really not concerned about the "now," as much as it is the past and the future.

You can think of consciousness as a director of a movie, whereby the director is providing the descriptive framework of the actions, movements, delivery, and general flow of the script, but is not in direct control of the actors and the equipment themselves.

This is not to say that a director cannot achieve the desired net results, but it does suggest that those results often come by way of future scene takes and are the results of coaching, the application of experience, changes in scenery, camera angles, lighting, and such.

In a sense, the director determines the outcome, but is dependent on many other things working together in a congruent way. The director, almost always, is making changes to historical data. This could include something such as suggesting acting changes from what was attempted in the last scene take, or improving poor camera angles when reviewed on tape. The director is then changing the future, in a sense, by guiding and altering the next attempt.

The conscious mind operates in a very similar way. Although it is not making the decisions of the present moment, it did conceive of the direction, laid down the constraints, and is monitoring the progress. And just like the director, changes can be made, but they will be made in a future tense after adjustments are in place.

In addition to the objective data points, like an actor sneezing, there are also metadata that can be gathered from the last take that will allow the director to further fine tune and/or make decisions to maximize the next take. In this hypothetical case, the director realizes that the actor has been sneezing all morning and subjectively determines that the he probably has a cold. The director may then use this metadata to assist in his decision to halt production for the day in order to give the actor a chance to recuperate. The connection or linking of subjective metadata to objective data becomes even more important as we continue our examination of brain function.

Just like the director or an advanced database, the brain applies metadata, or contextual information related to the primary data it is processing and storing. Metadata is ancillary data related to the primary data, or simply data about the data. For example, your brain may store, as a primary data point, the scenario, actions and reactions to touching a hot stovetop. Metadata connected to that event would be registering of pain, and hopefully a true/false bit would be set, a "flag," if you will, to mark this as a negative event.

After an event happens, the conscious mind parses, reviews, and analyzes the data. It is during this process that the brain compares the current data with historical data and the associated historical metadata. The final step is for the conscious mind to assign a new metadata "entry" to the current event. This is where the objective data is assigned subjective value judgments such as "good" or "bad," "hot" or "cold," "pleasure" or "pain," and so on.

We should note that the origination of these subjective values, "bad" for example, is derived from a prior learning process.  In the stovetop example, as a young child we may not know what the word "bad" means, but the burn that originated from the hand on the stove is remembered as a negative incident. Later, as our cognitive reasoning expands and we continue learning, we realize that the word "bad" is generally associated with examples such as this.

It could easily be argued that the aggregate of all metadata contained in the brain is where emotion, values, and morality stems from.  The reason for this is because the primary data elements themselves are only objective data, where the metadata is capturing and applying the subjective rationale associated with that data.

As mentioned earlier, new incoming data must pass a filter threshold of "noteworthy" or significantly different from prior data, in order to be consciously given bandwidth for processing.  For example, as you drive your car across town, you are presented with hundreds or thousands of terabytes of information; noticing grass growing, wind blowing the trees, the passing of other cars, the sound of the car's tires on the road, birds flying, insects on the windshield, and a myriad of other things.

Your brain, at some point in your past, has already studied and analyzed each of these items. And unless there is something significantly different about anything you see, hear, smell, or feel, your drive will end up being "inconsequential" from the standpoint of learning, and therefore your conscious mind will not pay close attention to any of those data inputs.

It is also interesting to note that as you drive your car, you do not need to pay much attention to the actual operation of the vehicle. This is because back when you were learning to drive, your conscious mind has already analyzed, understood, and coordinated all of the motor skills, rules, and reactions necessary. Your subconscious mind is therefore doing the actual driving, and is operating on the training data and experience that your conscious mind has already processed and stored. In just like every other part of your life, the subconscious is making all of the decisions and running the show, moment-to-moment.

So if your subconscious mind is driving the car, what is your conscious mind doing during this time? It is analyzing and processing dynamic incoming data; filtering, storing, analyzing and feeding that newly arrived relevant historical information to the subconscious mind so that it can be used during the process.

Remember that although the conscious and subconscious serve different roles, it does not mean that they work independently. Just as the director takes an active role in coaching actors, the conscious mind is continually analyzing and evaluating new data, and then updating the data or instructions, and constraints that the subconscious mind can use to carry out the process.

*Consciousness is the director, organizer, and manager of your life. While the conscious mind does not make decisions of the moment, it lays out the parameters, constraints, and framework for those decisions to be made.*

# Data Filtering

As mentioned before, the brain is taking in an enormous amount of new data every minute, and to keep from literally being overloaded, it must filter the data based on "newness" or data that is in some way unique from what has been analyzed and understood in the past. This filtering process is automatic, and is not something we must consciously put in place.

This newness of data begins at the moment of birth but becomes very pronounced when children are old enough to move about on their own. From that point forward, they are in exploratory mode soaking up all available information.

It is precisely this reason why young children can be fascinated with ants for hours, or playing with simple toys can take up so much of their time. They are experiencing and learning these things for the first time. This is also the reason why the passage of time seems much slower when we are children as opposed to when we are older. The newness of the world and the focus we apply to studying and learning each and every thing we come across causes time to appear to move more slowly. The older we get, the less "new" we are exposed to, and consequently the faster time appears to move. Later in life, time dilation or slowing occurs less and seems to be most obvious when we are in the middle of something like an accident. It is the extreme focus that our brain is applying to analyzing newly arriving data in these situations, that we experience a "slow motion" effect. Time itself isn't actually slowing, but rather our perception of time is altered because of the focus and deliberate processing our brain is performing on as many data points as possible [3]. Any event in which time dilation occurs represents the closest example of our conscious and subconscious working in unison.

Our mind also seems to be inherently driven towards acquiring new data. This is one of the hidden motivators for exploring new things. As children play with toys, for example, and once they have "figured out" how something works, they usually have a diminished interest in that item going forward, and will move on to exploring new things. All of this is a natural progression of learning.

The mind's data acquisition process repeats itself continuously every moment of every day throughout our lives. Our brains, necessarily as adults, must filter and work on an exception basis in order to have any sense of focus for accomplishing tasks. Knowing how this focus and learning process works, it should seem ironic that adults become frustrated with a child's lack of attention when we are trying to get them to do something, or how easily they are distracted. Instead, this should be understood as natural data gathering and processing. Focus and attention will come later as smaller quantities of newness are speeding into the brain.

*Data filtering is crucial for the proper operation of the brain by limiting and controlling what the mind must consciously parse and contemplate. The amount of newness arriving to our brains also affects our perception of time.*

# The Power of Metadata

Throughout our lives, subjective metadata is applied to all of the objective data points we acquire. For instance, the hot stove example earlier applies subjective judgment metadata that tells us fire is hot, and burns hurt. Later, we link this same metadata with other related items like logs burning or hot engine exhaust pipes. By applying experiential and subjective metadata to new incoming event data, we form, over time, traits, values and wisdom.

And not all metadata need come from hot stoves or fire ant bites. Children and adults both have a propensity to also take directional cues from authority figures, especially as children, and incorporate those as value-based metadata as well.

If we grow up in a Buddhist household, and are regularly exposed to those beliefs and practices, we are likely to maintain Buddhist beliefs going forward. This holds true for not just religious beliefs, but just about every other belief and value type that humans demonstrate. Although our values and beliefs are often created through exposure, they can, and often do change based on new ideas, exposure to others, and personal growth over the years. Humans are not unique in the animal kingdom with offspring learning from and mirroring their parents. That behavior is simply another biological predisposition required for survival.

Metadata necessarily becomes the foundation of emotion because it is not an empirical or objective data element. In our hot stovetop example, the stove, the burner, and the heat generated by its operation are all objective data points. The metadata is the applied reaction: pain. In other words, metadata is the description of the results of the event, not the event itself.

Understanding metadata, and how our brains link that metadata to the empirical and objective data that we acquire through living, makes it easier to understand how emotions evolve. We can clearly see their origins by using a straightforward scientific approach. Methodological reductionism [4] is a process by which complex systems are understood and explained by breaking them down into smaller and smaller parts. By analyzing and understanding those parts individually, we are then able to understand, what at first, was a confusing and complex system.

By using this approach, reductionism can get to the root of most emotion and demonstrate that they originate from basic attributes of pain, pleasure, and other reactionary responses. The summary of groups of this metadata in each "bucket," pleasure for example, linked with exemplary experiential data points, form the basis and threads of emotions, values, and beliefs.

Once something new is experienced, it is this aggregation and linking of the metadata that allows emotional growth to occur. This is especially true for linked experiential metadata, which in and of itself, and regardless of the topic, could be viewed as the definition of wisdom.

And we need to remember that metadata is not rationalized data. It is subjective data tied at the lowest level to basic functional responses such as pain in our stovetop example. Pain equals bad. Bad is interpreted to mean avoid. Avoid builds to fear, and so on [5].

Another source of persuasive impact on metadata is through social networks, and is especially true during our formative years. Just take a look at a teenager in high school and try to estimate how important it is for them to "fit in." Since, on average, teenagers have yet to accomplish much in life, their sense of self worth is almost entirely tied to superficial things such as their looks, friends, social standing, and extracurricular activities. The "values" they create during these years become the benchmark metadata they apply to many of the future events in their life.

And without being consciously aware of their impact, many of the "values" created during those years will actually end up becoming constraints later in life. For example, many people at the ages of forty, fifty, or beyond are still "controlled" to some degree by social value metadata they incorporated during the years of middle and high school.

The impacts of socially derived metadata don't end with high school graduation, however. They continue to impact our thoughts, decisions, values, desires, and beliefs throughout our lives. The herd mentality and "group think" is pervasive to human behavior.

Purging, reevaluating, or even possessing the willingness to reexamine old "bedrock" data and its associated metadata like this is an obstacle many adults never overcome. The "stickiness" of early-acquired data seems to be well cemented into the brains of most people.

What becomes clear is that metadata, being reactionary, low-level response data, is not subject to direct conscious rational thought. And in case you've ever been in an argument before, you will know that it is therefore an exercise in futility to directly challenge the emotions, values, and beliefs of someone else. This issue occurs due to a fundamental difference between data types. Objective data, of the type the rational mind can work with, can be consciously reviewed and examined, and basic metadata, used by the subconscious, is largely used for automated modulation of the system as a whole.

These differences will become even more apparent later as we discuss implementing changes in views, opinions, beliefs, and free will.

> *Metadata is data about the data. this type of data is used to supplement objective data by applying context and capturing low-level reactionary data. Almost entirely subjective, metadata is immune from rational contemplation and is the origin of emotion, views, opinions, and beliefs.*

# Subconscious Operating System

Rather than reviewing and using specific experiential objective data points like the conscious mind, our subconscious uses the aggregation of the metadata and highly summarized non-specific objective data as a basic operating system.

Again, it is important to understand that in the case of a hot stovetop, your higher brain regions and the conscious mind that resides there was not involved in the rapid removal of your hand from the stove. This reflex action is controlled at the subconscious level, and we should be thankful that it is, otherwise the delay for the conscious mind to process and make the decision to remove your hand would cause further injury than the "automatic" response.

Because of these things, the subconscious mind can be thought of, more or less, as the "kernel" or core of the operating system of your mind. Involuntary bodily functions such as breathing, and regulation of systems happen automatically in this realm. Layered on top of these basic functions, in the same subconscious region, are the instinctual reactionary controls, many of which are the aggregate of metadata as outlined above. Understanding this hierarchy of control, and the effects it has on the body, makes it clear that these functions are a biological imperative for survival and necessitates that your brain operate in this way.

But the realm of control for this part of the mind goes well beyond just rudimentary automatic system functions and reflex actions. The subconscious mind makes all of your decisions; from the choice of clothes you wore today, to the food you ate for lunch, and the value judgments you made about the people you passed on the street.

The subconscious mind is the engine running your daily functioning. All of those random thoughts that "pop" into your conscious mind were simple telegrams from your subconscious. The subconscious is also that annoying DJ that will not quit playing some old song that's stuck in your head.

Why do things like these happen? Usually they happen by way of triggers. In the case of a song, you could simply hear a phrase during a conversation, or think of an event from the past. Your conscious mind then makes a quick query on historical data, and says, "Oh, yeah, I remember..." and that's all it takes to turn the subconscious DJ loose with disks spinning. And sometimes it seems that DJ is wearing headphones, singingly loudly, and completely ignoring the conscious mind telling it to "Turn it off!"

And this works with not just songs, but with all

of your thoughts. Triggers are everywhere... in the news, conversations, something you read or see, a sound you hear; any incoming data is subject to becoming a trigger for a query and subsequent contemplation of the results.

Triggers are the primary starting point for branching the subconscious processing. They are the basic on/off, true/false "flags" used in the operating system of the mind to keep the "engine" aware of current changes in the environment.

These flags are used during conversations. For example, someone may be talking about a problem they are experiencing with their car, and, in parsing that sentence the conscious brain retrieves data that reminds you that last week you noticed your tires were wearing and needed to be replaced. In this way, the conscious mind is simply retrieving related historical data.

What you say during conversations is directed by the conscious mind, using historical data, but physically created and delivered by the subconscious. The conscious mind is querying data, building an idea, and connecting relevant information. The conscious mind, however, does not choose the specific words or syntax. It merely delivers the summarized idea to the subconscious.

These ideas are then used as a new temporary thread in the subconscious that uses prior learned data and metadata to choose the best words to convey the message, organize syntax, regulate the volume and tonality appropriate to the context and setting, and also coordinates your breathing, diaphragm, lips, tongue, and facial and body movements. During the delivery, your eyes and ears are taking in data from you, your listener, and other potential audience members.

The conscious mind filters and stores this information and then uses it to make adjustments. It may, for example, tell the subconscious to "lower the volume" based on looks or gestures seen from others in the room. It may detect a look of disagreement on the face or in the body language of your listener, and tell the subconscious to "restate your point using this other data…" or simply "move on to another subject such as…"

Remember, the conscious mind is the director, and the subconscious is the actor. Although it may be more appropriate in this analogy to realize that the director is supplying basic ideas and facts, where the actor is more improvisational, using a loosely defined script based on a prior learned inventory of ideas, terms and conditions.

Triggers are, by definition, useful as a memory aid. Connecting a word or short series of words to a larger idea is the basic premise of taking and studying notes. Later, during a test, reading those key words in a question help the conscious mind retrieve the accurate results for the answer.

The same holds true for using shopping lists, to-do lists, and calendar entries. All of these things are simply lists of triggers for your conscious mind to use in order to "steer" the subconscious into action.

> *The subconscious operating system is where your thoughts arise and your decisions are made. This area of the brain can be thought of as the "engine" that automatically makes these decisions using a pre-constructed set of guidelines created by the conscious mind. The subconscious will also use dynamic data changes to telegraph thoughts up to your conscious mind for contemplation and further direction.*

# Decision Making

As can be seen, the process of making a decision really comes down to the data that is available to the subconscious mind at the moment a decision is needed. Since this happens below the level of consciousness, it appears to be "from nowhere" although in reality it is simply the result set supplied by our subconscious based on the specific query we just gave it.

This is not to say that changing our minds is impossible... quite the contrary. The conscious mind can, and often does, retrieve older data for the purpose of re-comparing it to something relatively new, beyond the first-pass comparison provided by the subconscious. That re-comparison and contemplation process allows us to "rethink" things, reevaluate situations, and fine-tune our decision making process. We may, in light of further scrutiny, decide to make a different choice, or, we may revert back to the original solution provided by our subconscious.

"Gut feelings" as we have all experienced, are simply first-pass results from the subconscious based on the event data we have at hand. In other words, dynamic input data from a current event will be used by the subconscious mind to make initial decisions. This process is very similar to using Google, in that when we are searching for something, Google will deliver a set of results based on whatever keywords we typed. If we are not completely satisfied with those results, we are free to alter, change, or add-to those keywords to further refine our search.

Given that most events are not exactly similar, the initial responses from the subconscious will be compared to historical data (rationale) and historical metadata (emotion) in the conscious mind, in order to form follow-on reactions or decisions.

Although on the surface this might seem confusing or complex, it really is quite simple. The conscious mind is the Supreme Court, while the subconscious are the lower courts. In all cases, decisions are first made in the lower courts, and if the magnitude of the issue and complexity of the outcome warrant, initial decisions are moved upward for further contemplation. Unlike the US Court System, however, automated decisions in the subconscious are continually evaluated and adjusted by the conscious mind. In other words, most mental decisions are not "final," rather they are continually reviewed and reevaluated on the merits of new incoming data.

*Since the human mind is continually evolving, and newer data is always arriving, the conscious mind can reassign new metadata to older ideas, which in turn have the ability to move thought patterns or values from one generalized "bucket" to another. The source of the new data, how that data can be rationalized with other older data, and the frequency and magnitude of similar supporting data play a large role in shifting mental positions on a topic.*

# Reviewing the Brain Data Process

Lets take a moment to review the basics of the brain system and its operation as have been outlined thus far.

- Through your nerves and sensory organs, your brain is constantly receiving new data.

- The data it receives can only be dealt with from a historical perspective. In other words, new incoming data must first be processed before it can be contemplated and utilized.

- A filtering system is applied to "weed out" unimportant, or insignificant data, based on newness or uniqueness.

- Data passing the basic filter is temporarily stored for processing.

- The conscious mind reviews, contemplates, and rationalizes about this temporary data, comparing it with available historical data and historical metadata.

- The mind then stores this data across

several regions of the brain, with higher or lower resolution in the regions that contributed more of the data inputs and/or applicability and magnitude towards a certain region.

- Finally, the mind stores a value-flagged, metadata "synopsis" of the event specifically for "run-time" use of subconscious operating system.

# Steering Free Will

Making decisions can be both deliberate, based on delayed contemplation, and automatic, based on subconscious. We learn, build values, morals, and wisdom by the connection or linking of data and metadata.

The automatic nature of our first-response results delivered by our subconscious is merely that... a first-response. Steering our free will really means that we have the ability to influence future subconscious responses by thinking.

Very few things of any consequence are going to be a simple binary decision. In almost all cases, choosing to switch from Buddhism, as in our earlier example, to Hinduism would require deliberate reexamination of several, possibly hundreds, of prior held beliefs, values, objective and supporting metadata points. In a case like this, the trigger or triggers for even beginning that contemplation would need to significant.

On smaller magnitude issues, the size and scope of the elements involved become more and more manageable. And as we will see, significant change is often the result of many prior smaller changes, rather than an abrupt, overnight epiphany, although that does sometimes occur. To grow, evolve, and learn, our conscious mind is continually analyzing newly arrived historical data. Unfortunately for many people, over time as they age, there appears to be a reversed learning curve or reduction in the desire for uptake of information that does not closely align with previously known data.

The data filter that we previously discussed is not static; rather it evolves in concert with learning. Consisting of aggregated metadata, portions of the filter are automatically and biologically entrenched, such as body function and regulation, where other portions are subjective, like values and beliefs.

The "strength" of the data filter is therefore directly connected to the aggregation of all previous subjective metadata and known summarized objective data, along with the "value" flags that were assigned to those metadata elements. In other words, throughout your life, your conscious mind assigned varying levels of value (high to low) on the metadata attributes it uses to describe the objective data.

This is why, for example, one person may find tattoos repulsive while another adores their ability to display self-expression. Neither is intrinsically right or wrong. It is likely that nothing more than the value flags in one were historically influenced strongly by their parents or peers.   And since metadata is reactionary subjective data, neither is likely to be consciously aware of the rationale behind their thoughts.

How could someone who initially finds tattoos repulsive evolve over time to being someone who has several?  This kind of shift in camp has occurred in millions of people on just as many topics.

The most likely trigger candidates for this type of change comes from data we receive from other people, especially those people whom we revere, hold in admiration, respect, or find interesting and challenging.

Social data is a strong driver in changing our internally held metadata attributes.  If all of our friends become enamored with tattoos, and not only get their own, but speak of their virtues, it is likely that over time we will begin to change our attitude towards them as well.  In this hypothetical example, as we witness the social bonding, and the seemingly uplift in their social strata for those "in the club," we may begin to feel left out.

> *The biological herd mentality and the need to "belong" to the group likely drive many of our low-lying metadata attributes and their ranking of significance as a value parameter in thought and decision-making.  Thinking outside the box, or simply thinking*

*for one's self has merits well beyond any short-term frictional change.*

# Desire

At its most basic form desire is the driver for change, and is conversely the strongest barrier. With it, there are almost limitless things we can accomplish. Without it, there seem to be unmanageable chasms to cross, endless barriers to overcome, and an almost premeditated will to fail.

The desire for education and experience is the starting gate. These are the nexus for success. If we have, or foster, a desire for newness, this drives the subconscious to telegraph requests to the conscious mind to seek, gather, and process new information, and ultimately, arguably most important, to coach the subconscious to take action.

As we move through the years, we have a tendency to become mentally sedentary. We change from a "conquer the world" approach to a "preserve what we have" attitude. If necessity is the mother of invention, then being comfortable must be the destroyer of desires.

The reality is that most fifty-year-olds are much less likely to take risks than a twenty-year-old. And these risks need not be entirely monetary. Seemingly valid risks could include changing your core beliefs on a certain subject, changing your lifestyle, starting a new career, or even deciding to get married or divorced. One could argue that this is simply because a fifty-year-old normally has more to lose than a twenty-year-old. Or, "There is more time to start over at twenty," they might say.

And while this might be true from a simple money or materialism viewpoint, it is very untrue from a wisdom and likeliness-to-succeed standpoint. This does not mean that someone in their twenties should not try to grow, navigate uncharted waters, and try different things, but it does suggest that someone in their fifties should have an advantage doing the same things, so long as they hang onto the "conquer the world" attitude of their younger self.

We should also note that continual mental growth necessitates the application of action to the equation. Learning information, for the sake of information alone, is not very valuable until it is applied in the real world. Put another way, learning without action is worthless. The world is replete with armchair philosophers, but is aching for examples of action and leadership. And the most fundamental example of being a leader is simply the one taking action.

Given that the human herd mentality exists, with each new "leader" that emerges, whether they are the starter of a local gardening club, a rapidly rising nationally-known politician, or anything in between, there seem to be a never ending supply of people who want and need something and someone to follow. The contention here is that it is not wrong to follow others, we all need change agents and mentors in our life, and however we should simultaneously retain and foster the desire and actions to be a leader ourselves.

*Desire is the primary driver behind learning. Continual education and heading in new directions often requires challenging your existing thoughts, values, and prior historical data, and many times those challenges require internal leadership qualities to push past barriers that currently hold us back.*

# Fear

Fear forms another set of aggregated subjective metadata, and is manifested as the close anti-cousin of desire. Metadata flagged as "fear" accumulates in many dark corners of our mental databases.

The primary issue with battling fear is that in many cases your mind uses emotional metadata connected with some amount of rationalized objective data, and constructs a "rational/ emotional" response. This means simply that there is enough rational data in the mix to warrant concern and prevent dismissal, and therefore can become strong enough to alter decisions and action.

It is important to remember that fear is learned, and therefore we also have the ability to mitigate, manage, or unlearn these same things. And we should never underestimate the power of fear. If we consciously walk through and examine the constraints and restrictions in our lives, in almost all cases we find that most of them stem from some form of fear. Anything this powerful and controlling deserves exploration, as mitigating or conquering fears opens the doors to limitless opportunities.

Confronting fear might seem like something

we do not want to do, however if you have ever done this, on any topic, and later look back, you will be astonished, perhaps even embarrassed at your former self for having been controlled by these hidden "demons." It could be argued that conquering fear is one of the most liberating things we can ever do in our lives.

No one likes being controlled by someone else. Imagine how ridiculous it is when you realize that allowing fear to remain in your life is literally self-imposed restriction. Fear is willfully constructing a box or cage that you choose to live within. Viewed from a rational perspective, we are forced to wonder why anyone would ever allow this to occur.

One could argue that the real value in living really comes down to just a few things; the relationships we build with others, and the learning and experiences we gain along our life's path. Dying at the age of forty after having lived a robust life with lots of experiences could be viewed as more valuable than living to the age of eighty with substantially less experiences due to the restrictions of fear.

One path might be "safe," from the perspective of staying within our self-constructed cages of fear, but in the end might well be reviewed as a long path of wasted or missed opportunities. If we believe that there could be a purpose to our life other than taking up space, allowing fear to exist and control our directions would surely be one of the first things we would want to eliminate. Fear will also temper our rational mind from abhorring it. In other words, we have a tendency to live with fear, even if we consciously know it exists, because we have convinced ourselves that our lives are going along just fine as-is. This is akin to the Stockholm syndrome where the hostages begin having positive feelings for the captors.

All of this is not to suggest that we should simply throw fear to the wind and take risks that might endanger our lives or the lives of others. Rather it is the premise that deliberately using rational thought and methodically confronting and eliminating fears could be one of the most important projects we ever undertake.

How do we begin? Let's take a hypothetical example case and study how fear controls us, and a very simplified example of how to approach the problem. First, let's say that, for reasons not immediately obvious, you have a fear of flying. We might be tempted to say that this fear trumps your ability to have any free will on the subject, and the decision never to fly is automatic, and therefore beyond your ability to change.

If you want to change your position on this topic, the key word being "want," you must first be willing to understand where the fear emanates from, and then take steps to study and review the drivers of that fear. I've heard more than one person say it's really a fear of death, and self-preservation takes over to eliminate possibilities, which include flying. And while that's true at a very basic level, it is not really the answer we need if change is desired.

By taking in new data points, in this case examining how airplanes are constructed, how they fly, their safety record compared with driving, and the training requirements for pilots, one can begin to expand their understanding of a topic like flying and through that understanding, along with increased familiarity of practice, have the ability to overcome those fears.

By educating ourselves, we are able to mitigate otherwise paralyzing fears. This is not to say that we will or should become fearless in everything. It merely suggests that mitigation of fear by education, familiarity, and the knowledge of probability will enable your subconscious mind to make future decisions differently that it would today.

As we see time and again, education is almost always the mechanism of how we overcome fear along with most other limitations in life, and "re-wire" our subconscious to deliver new results. Surprisingly, the majority benefit of "education" comes from within, rather than from external sources. This is not to say that taking college classes, or reading books is not educating for you. It does, however, mean that the real value of data acquisition, whether through books, classroom or experiential, comes from the focus and effort you place on the processing of the historical data, what you just read or did, and devoting the proper attention and focus to analysis afterwards.

Time must be spent consciously linking new data to old data. In other words, the value of education is really about how you link and relate what you just learned, to everything you've already learned. Without the application of data relationships, learning is nothing more than gathering facts, which might make you unbeatable at Trivial Pursuit, but will do little for your ability to grow and succeed in the real world. By linking new data with old data, both objective and meta, we build a broad set of useful information and skills upon which to draw from. Only then does intellectual synergy, which should be the ultimate goal of education, become possible.

> *Emotional response often overpowers rational thought, which is why fear exists and also why we have such difficulty confronting it. Intellectual strength and synergy is the best weapon to conquer fear.*

# Intellectual Evolution

In order to "gain control of" or "steer" your free will, there is first a requirement for time dedicated to thinking. And it is the suggestion of the thesis of this book that time is dedicated for the rest of your life. In other words, the process of contemplation and revisiting old ideas, values, and concepts is not a one-time exercise.

Time should be set aside each day for thinking, through meditation, walks, sitting on the swing, or whatever method works best for you. The point is that time is needed to mentally raise and address issues, pick subjects that have troubled you in the past, or subjects that have been an area of interest or intrigue, and then formulate a game plan to find and gather new information on the subject.

To overcome inertia, we must place value on change, and be willing to dedicate time, even if we think we have no time to spare. Because our subconscious rank orders everything we do on a priority basis, if you want the time to study and think, your conscious mind must rationalize why this is important and direct your subconscious to allocate whatever time is necessary. This may need to be a daily conscious effort for a month or two until it becomes a standing habit.

Intellectual evolution as a concept should be straightforward and easy to understand; yet for many it is a daunting task in the beginning. Start with the idea that your brain is already hardwired for learning. It craves it. This is the process that our brain automatically employed as a young child. And remember that as we grow older, and the arrival of newness slows, we have a tendency to become mentally sedentary. Instead of hungering for new knowledge, we often come home from work, tired and mentally spent, and choose to instead sit in front of the television. If we are already doing this, we have just identified all the time we need for thinking and mental contemplation. We only need to substitute one for the other.

The imperative for making a conscious effort towards intellectual evolution becomes self evident as we are immersed in the path. But in the beginning, we may need coaching, support, and stimulus from like-minded friends and acquaintances. Surrounding ourselves with smart, open-minded and forward-thinking friends is essential for rapid growth and development. These people will be the provider of new ideas, and will be an excellent sounding board for us to bounce our ideas.

If you are already surrounded with these types of people, consider yourself lucky and continue down your path. But if we find ourselves in an intellectual vacuum, it would be wise to seek out and build some new relationships to help balance our existing life with the one we want to build. Integrating yourself into, or building a new social network comprised of these types of people is not difficult, but does require work and dedication.

A great place to start may be with groups such as TED or Technology, Entertainment, Design. While the name of this particular group may suggest a rather narrow focus, the speakers cover topics as diverse as world hunger, religion, farming, and virtually every topic imaginable. TED is a global organization, and as such the featured speakers are selected from around the world. TED Talks can be viewed online and TED Conferences can be attended twice per year.

TED also has localized events known as TEDx. These events can be tracked via the TED website, and if you do not see an event in your area, the system is designed for you to be able to coordinate and organize your own local TEDx event.

There are many other local and national groups and organizations flourishing with smart, open-minded and very personable individuals. These types of things, in any of their forms, are excellent networking opportunities and will put you in direct contact with intellectuals in your area.

One point to remember is that we may sometimes not think of ourselves as intellectuals, when in fact we are. Everyone possesses some level of knowledge that others do not. And even though organizations such as TED may offer stimulating and thought provoking discussions, we should keep in mind that the best source of learning may well be from ordinary people all around us.

We should never make value judgments about someone, no matter how we might at first perceive them. A friendly "hello" and a short conversation not only has the opportunity to brighten our day, but we never know when a stranger may say something that triggers new ideas or thoughts, and those ideas or thoughts could have the power to significantly change our direction.

Dismissing others, regardless of their social standing or appearance could very well cause us to miss some of the biggest opportunities in our life. One of the primary missions of intellectual evolution should be both to share and gather information with others. We should never let our current self-opinion hold us back from pursuing and forming new relationships with others. By making a conscious decision to interact with new and interesting people, no matter where on the curve we are today, we will almost guarantee to improve our knowledge and capabilities.

Beyond connecting directly with people, our

intellectual evolution will also entail seeking knowledge from traditional sources such as books, classes, instructional videos, workshops, and hobbies. These activities will be both the initiator of information you can share with others, as well as the "homework" you perform when exploring the ideas you gain from others.

> *Intellectual evolution is the process of evolving, growing, and expanding the mind. One of the best ways of initiating this process is through new relationships and networks of friends and acquaintances. Traditional learning methods such as books, classes, and other sources are utilized for depth of knowledge.*

# Thinking Time

When you carve out the time and begin your thinking, it is important to first spend some time letting your mind clear from the noise of the day. Put the bills, work, kids, house, and other distractions to the side for a few minutes and allow yourself to focus on an idea.

Exposure to new ideas is important, and this comes primarily from other people and experiences. If you spend every day doing basically the same things, with the same people, it is highly unlikely that you will be exposed to many significantly different viewpoints and ideas beyond those that you already hold. Growth, and the opportunity for growth, will be significantly reduced.

Being in control of your destiny in terms of decision-making, and steering your free will requires nothing more than the willingness to be open-minded and a desire to grow or be different.

You may be asking yourself "what does open-mindedness have to do with free will?" The connection is simply one of change, and the willingness to change. Often times this means moving outside of comfort into unfamiliar areas of thought.

The other, arguably bigger issue that many face is the question of "why?" Why should I even consider reviewing my foundational beliefs? Why would I want to learn to fly an airplane? What good is it for me to reconsider my opinion of illegal immigrants? And on, and on. This, above everything else, seems to be the primer for intellectual advancement.

A great number of people seem to lack the desire to challenge anything they already "know." In these cases, the metadata is emotionally walled off and will not be willingly exposed to rational scrutiny. Whether this is present or not is the most basic definition of open- or closed-mindedness.

It appears, however, that even the most sacred of personal beliefs or values can change, if the right data is available and a sufficient support community is in place and it becomes socially advantageous to change ones opinion about something or conversely, becomes socially stigmatized for hanging on to a certain belief.

Edward Bernays' principals outlined in his seminal book on the topic of Propaganda from 1928 were, and still are in use today to influence public opinion on everything from which shampoo to purchase to patriotic alignment during wartime. One thing becomes obvious from a higher-level observation of society, and that is, as we stated earlier, most people tend to follow, more than they individually lead. And, also as was said earlier, most people are hungry to be led, even, sometimes, if the direction is not congruent with rational thought.

> *Utilizing value-based metadata gleaned from peers and groups that one identifies with enables the herd mentality. Being aware of these factors, and consciously removing them can give you the freedom of latitude to grow.*

# Barriers

There is a process in the brain, during the conscious historical data review, whereby groupthink is sometimes allowed to have higher value than personally acquired metadata. Argumentum ad populum, a Latin phrase for "appeal or appealing to the people," is often used when referring to this phenomenon, but is best described as the concept that "if many people believe something, it must be true."

This influence can be driven from a number of personal fears, insecurities, or other social factors, but the net result is that sometimes in the data filtering process, a person rationalizes allowing someone else to think for them. Most likely a very old biologic leftover, groupthink, which was likely useful originally as an instinctual safety mechanism, can be a very dangerous and debilitating barrier for cognitive rational growth.

Coupled with metadata obtained from early influences, groupthink can be a powerful aphrodisiac luring the conscious mind into a false sense of knowledge and security. It appears that many times it becomes easier to go along with groupthink, rather than spending the mental energy to analyze, rationalize, and possibly divert from the herd. And in some cases, many may continue with the herd, even if they have reached conclusions contrary to popular belief. Perhaps the cost-benefit analysis of declaring independence does not outweigh establishing a new set of relationships.

Other enablers related to groupthink can sometimes be rationalized on an individual basis by using a partial-truth justification. In this scenario, someone may publically go along with groupthink, while privately entertaining doubts and questions. This seems to be a common theme during election years, where someone will vote along party lines, even though they may have doubts about proposed plans discussed during the election campaign.

The partial-truth justification works for some by admitting portions of an idea are most likely incorrect, while willfully neglecting to apply rational logic and scrutiny to the remainder. The "truth" portion of the partial-truth justification may not be true at all, but in order to avoid confronting fear, or to remain in good standing with the herd, the conscious mind is walled off from doing its job. Alternately, this could the thought of as the "lesser of two evils" principle, with the lesser being going along with the herd.

The idea of possibly being wrong is yet another barrier to questioning and reviewing values and beliefs. Admitting that we may be currently holding an outdated or incorrect viewpoint is, to some, an admission of ignorance. This is why we rarely hear someone admit they are wrong during an argument, even if they begin to realize they are.

Yet another common barrier is simply that many people do not really know why they believe something to be true or false, only that they do. Maybe this is something they were told as a child, or heard from others during their formative years. Perhaps it is just that their position on the topic has been held for so long that they assume it must be correct.

This particular barrier is especially tricky because, again, no one wants to confront the fact that there is little rational basis for his or her thoughts. Arguments, whether internally or with others, in this arena, like most of the previous examples, are by definition emotionally based, rather than rational. The subconscious will argue with metadata rather than being directed with rational objective data from the conscious mind.

Lastly, even if someone becomes willing to contemplate new viewpoints, many times they are faced with cognitive bias. This means, simply, that they will have difficulty objectively analyzing an alternate viewpoint without apply their existing beliefs. This bias is a powerful barrier that adds a significant amount of friction to change, and therefore can, if unchecked, many times render the exercise pointless.

*A better approach to overcoming all of these barriers may be to view discovery and change as a superior position of growth and existence. Rather than entertaining admission of prior ignorance, perhaps standing proud of growing in new directions, is a better framework for personal growth and success. These "going out on a limb" directions, even if others do not immediately follow, are the fundamental basis of leadership.*

# Compartmentalization

As mentioned in the beginning of the book, the value of data is significantly increased when it can be contextually linked to other relevant data. This not only increases the scope of the information, but also allows us to relate it to a much wider context of knowledge. Doing this can provide a litmus test for basic truths.

Compartmentalization, on the other hand, has a tendency to dilute truth, knowledge, and focus because of segregation. Simply defined, it is the process whereby we keep ideas, values, beliefs or behaviors separated in our minds in order to avoid the conflict they mentally impose. It is, in effect, a defense mechanism [6] to hide behind when ones values and beliefs are not entirely congruent, or aligned with rational thought.

These basic struggles are often used with character development in books and movies. The "hero" almost always has character flaws and it is left up to the audience to weigh the good against the bad. And unlike the real people we interact with in our lives, on the screen, the audience is able to watch the internal struggles as the "hero" tries to balance their bad behavior against the "good" they are doing.

Compartmentalization extends well beyond simple daily thoughts. Many people seem to live highly compartmentalized lives. They may take on one role or set of behaviors at work, another at happy hour, yet another at the gym, and finally another at home. The segregation of values and beliefs in the mind is for many an escape mechanism from the reality of one's life.

Allowing compartmentalization to remain in place is akin to living with multiple personalities. As a person grows intellectually, it will become harder and harder to keep those areas separated and reconciled.

Instead of multitasking between personalities, we should recognize that focus and unity is much more powerful. The reason why a laser has the ability to create a hologram, and why it is used in so many other applications from surgery to DVD players, is because of coherence. Coherent light simply means being in alignment. In a broader sense it means acting in unison and moving in the same direction. Regular incandescent light cannot achieve these things because it is incoherent, or moving in multiple directions, therefore its power and effectiveness is reduced.

If we are motivated to have better control of our thoughts, and our control of free will, mental unification is required. A committee of incongruent ideas, values, behaviors, and beliefs cannot accomplish intellectual maturity or control.

*Rather than compartmentalization, a better approach is to unify your thoughts and actions onto a common front. Consistency, focus, coherence of thought should be a tool against compartmentalization. By eliminating these segregated areas of the mind, a much greater intellectual strength can be formed, and the ability to direct free will be greatly enhanced.*

# Honesty

Being in control of your free will necessitates first being honest with yourself. And we need to be sure to define honesty without grey areas. There may be multiple scenarios where one thing applies versus the other, but there are not multiple versions of the truth.

Being honest should be the foundation upon which all of our data and metadata is judged, stored, and used. Without honesty, both internally and externally, we really have no foundation for accomplishing anything of value. It should be obvious that no meaningful relationship can be built upon lies, just as no meaningful intellect can be build upon false data.

If we choose to build meaningful intellect, we must regularly and consciously shine the light of reason across all of our values and beliefs to ensure they are congruent with the rest of what we know and have learned. In other words, we must be ever vigilant to root out and remove any data points that fail to meet basic rational scrutiny. Anything that does not easily fit within this scope is probably false and needs to be reexamined.

Let's take a moment and look at why this approach must be applied to everything in order for our world to work as well as it does. The next time you are about to board an airplane and fly across the country, stop for a moment and consider the human achievement that makes this possible. The airplane you are about to board is the product of hundreds of years of discovery, thought, ingenuity, and hard work by tens of thousands of people. From metallurgy to electronics, chemical engineering to aerodynamics, computer programming to physics, large portions of many scientific disciplines are culminated in that one machine.

We must also not forget that the airplane itself is sitting on engineered pavement adjoining advanced building materials and construction techniques of the terminal. The airspace above is controlled and sequenced by computer, telemetry and radar systems. And ultimately you will arrive at your destination with the assistance of radio navigational and global positioning system aids.

In this scenario, one of thousands of examples in our modern daily world, there are literally millions of pieces and parts, one built upon the other, that are brought together to make this possible. All of these things are the results of science and the scientific method employed in learning and applying one theory to another.

How safe would we feel in boarding this airplane if

we knew that several areas of science underlying all of this technology were not factually sound?

From the plastic extrusions in the doorframe to the metallurgy of the engine turbines, every piece and part demands truth. And without it, your flight will likely not end well. Choosing to be dishonest with ourselves, even at the most basic levels of thought and mental compartmentalization, puts all of us at a disadvantage. Thinking rationally requires honesty, just as safe airplane parts require honest science. There can be no exceptions if we want safety in the air, and there should be no mental exceptions if we want intellectual excellence. Further, if we can't be honest with ourselves, there can be no expectation of being honest with anyone else.

This is one of literally thousands of everyday examples that underscore the need for honesty. The information, ideas, thoughts, values, and beliefs held in our minds are not only a benchmark measurement of our intelligence, but also provide a reference as to how much we have grown since childhood. Without honesty, we build a false existence in a fairytale. Given that our minds are our most valuable possessions, we should continually monitor and ensure they are not filled with, nor controlled by erroneous, ill-conceived, or outdated data.

*The path to intellectual growth is paved*

*with honesty. We must be honest with ourselves in order to properly filter, analyze, process and store the data that we receive throughout our life. If we allow our mind to entertain non-rational thoughts or ideas, or to believe unsubstantiated data, we run the risk of compartmentalization, or worse, not being in control of our free will.*

# Strategic thinking

Now that we know the basic mechanisms of how our mind works, and how to consciously deal with the many barriers and issues working against rational thought, we should the examine the tools for control and direction.

While we all possess varying natural ability, the art of strategic thinking is really a learned skill. As the name implies, strategic thinking can be thought of as strategy thinking. Its concern is moving above the tactical and observing things from a higher level. Simultaneously it can imply charting a new course, or operating on a wider or expanded scale.

Strategic thinking involves honing your abilities to see beyond the obvious, and to seek options, opportunities, and possible course changes necessary to take you from your current position to where you want to go. As an example, playing chess involves strategic thinking in that the players analyze the board and try to anticipate future moves by their opponent along with planning their own strategies several moves ahead.

Conversely, when we operate at a tactical level, there can be times that we are so immersed in the "noise" of current issues and tasks that we are unable to "see the forest for the trees." We have all been there. The juggling of work, domestic chores, kids, and everything else, can often cause us to become lost in our quest for direction, or even worse, after long periods of time, even lose our sense of purpose.

To think strategically, and to chart a course out of the tactical noise, we sometimes need to mentally climb one of those trees in that forest and take a look around. It is from this "higher" vantage point that we can consciously assess the situation at hand, and form a plan or roadmap for moving from where we are, to where we want to be.

From our previous example, it is interesting to note that playing chess is done from above the board so that the entire field of play is in view. Imaging the success of trying to play chess from the vantage point of one of the pieces. This is the difference tactical and strategic.

Part of the Thinking Time that we discussed earlier should be used for this strategic thinking exercise. It should be part of our daily regimen. By making time to practice analyzing the various historical data that we possess, our fears, our values and beliefs can be finely tuned to match the rational thought we apply to our lives going forward.

Strategic thinking should be the one of our primary tools. It should not only help guide our direction, but also to rationalize the connections we make between our historical data. We can use this process to purge old erroneous data, direct the subconscious to take action towards obtaining newly understood data deficiencies, and to enhance the understanding of data links by viewing their connections in a broader context.

*Strategic thinking is essential for planning and constructing a roadmap for success. By deliberately using thinking time for reviewing and planning, strategies can be formed to direct our subconscious to take action in gathering new data, and will enhance our existing knowledge by connecting information together in a broad context. Growing intellectually from where we are today will require deliberate observation and removal of the mental barriers and constraints that have held us back in the past, and strategic thinking can be the toolset for building the roadmaps and strategies that navigate and conquer our mental barriers.*

# Final Thoughts

By consciously adjusting our thoughts, we have the ability to influence the decisions our subconscious mind will make during its "automatic" operation. These adjustments can come in many forms, from simple changes in clothing preferences to major changes in values and beliefs.

Free will is not something out of our control. We have already directed and laid down the construct of how our decisions are being made, and we also have the ability to steer ourselves in new directions simply by rethinking exiting data, as well as deliberately gathering new.

We should not feel powerless by forces at work in our minds. Fear, anxiety, insecurities, and many other constraints can be dealt with and mitigated or eliminated by simply the application of rational thought and gathering new supporting data for a change in direction.

Although the human mind is complex and only partially understood, even with what we know today we are in awe of its potential. Unlocking the mysteries of the mind will allow us to achieve ever higher states of consciousness, and through open-mindedness and the application of rational thought, we have the potential to not only change our view of the world, but to actually change the reality of the world itself. Far from puppets under the control of someone or something else, the human mind is a powerful tool that can be sharpened throughout our lives. The most amazing aspect, and perhaps the most basic requirement, is all that is required is simply the desire to change. And if we possess that desire, there are limitless opportunities ahead.

# Acknowledgements

I would like to thank my best friend, wife and editor, Sharon Allard, for her contributions to this book, and to my life.    Sharon is also the CEO of our company, and has been a true partner in my quest for intellectual growth and learning.  This book is really the product of our conversations over the years, and her thoughts and ideas have been fundamental in setting the direction which led to many of the conclusions and theories that I presented in this book.

I would also like to thank my wonderful family and friends for always being supportive, even though I am sure there have many times over the years where they wondered what I would be doing next.

# About the Author

Adam Allard is Founder, President, and Chief Scientist of Allard Energy, Inc., a company that specialzes in producing ethanol fuel from waste products. Mr. Allard is an autodidactic visionary and inventor, and directs the research and development of fuel production equipment and software systems within the company.

Prior to founding his company, Mr. Allard spent many years in the information technology field, holding various managment and development positions. The last position held prior to forming his own company was Chief Architect of Information Technology where he also led Strategic Planning with an international Fortune 500 company.

Mr Allard also had prior careers building medical endoscopy equipment, designing systems for commercial FM radio broadcasting, oil and gas drafting services, and building training models for the US Department of Defense and US Navy. He has also been a guitarist, and a radio announcer.

Mr. Allard is a private pilot and enjoys building and flying airplanes. He is an active member of the Aircraft Owners and Pilots Association and the Experimental Aircraft Association. His other hobbies include scientific research, designing and building prototypes of new equipment, reading, weather analysis, and writing. He is also an FCC Amateur Radio Operater and has participated with the National Weather Service as a Storm Spotter.

Mr. Allard was also trained as a firefighter and was actively involved in both fire and medical rescue operations. He was later President of his town's volunteer Fire Department.

Starting in 2002, Mr. Allard spent eighteen months of evenings and weekends building his house himself. He regards this as a significant learning experience, although he has retired from the hobby.

Mr. Allard is married and has three children. The family lives in a rural country setting on twenty acres.

# References and Notes

1. Holonomic Brain Theory. Karl Pribram in collaboration with physicist David Bohm.

2. Ervin Laszlo, "In Defense of Intuition: Exploring the Physical Foundations of Spontaneous Apprehension," Journal of Scientific Exploration, 2009, Volume 23.

3. van Wassenhove V., et al. (2008) Distortions of Subjective Time Perception Within and Across Senses. PLoS ONE 3: e1437. DOI: http://www.plosone.org/article/info:doi/10.1371/journal.pone.0001437
Tse, P.U., et al. (2004). Attention and the subjective expansion of time. Percept. Psychophys. 66: 1171-1189.

4. Methodological reductionism is the concept that complex systems can be reduced and understood from the analysis of its parts.

5. Schacter, Daniel L. (2011). Psychology Second Edition. 41 Madison Avenue, New York, NY 10010: Worth Publishers. p. 310. ISBN 978-1-4292-3719-2.

6. Tangney. Leary, Mark R. Leary and Price, June, ed. Handbook of self and identity. Guilford Press. pp. 58–

61. ISBN 978-1-4625-0305-6.